PRESCHOOL MATH AT HOME

SIMPLE MATH FOR TODDLERS

Let's get counting, tracing and learning numbers for 4 year old

Rosi & Coni

Blue Cherries, Publishers

Enjoy our others workbooks of the serie

Conscious Teaching:

Available in Spanish and Portuguese
Available in several countries in America and Europe

Fullname

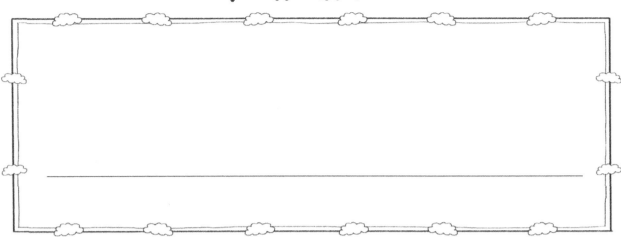

Date

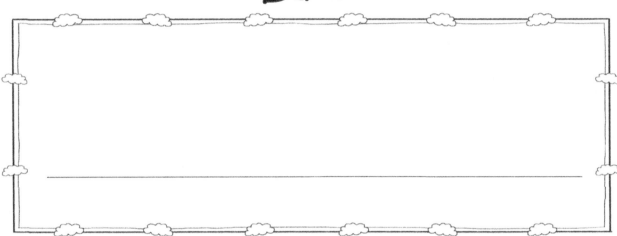

School's Name

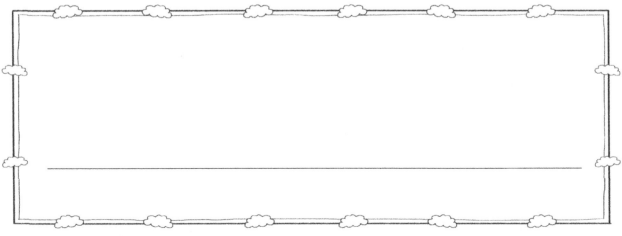

Dedicated to all children from 4 to 5 years of age, so that they explore the world of numbers in a fun way, acquire knowledge to use as tools in their day-to-day life and feel able of achieving numerical objectives with success.

When you buy this book you will receive a special gift!!

*If you have comments or questions you can write an email to **bluecherriespublishers@gmail.com***

Index

This book is composed of 8 parts in which the child will learn in a funny and a dynamic way the numbers from 1- 30.

Introduction

Preschool Math at Home welcomes you to the wonderful numerical world.

In our amazing trip the child will learn to recognize the numbers from 1 to 30, count and follow the numerical sequences, as well as, to associate the numbers with quantities and write them in a correct way. We advice you to do one activity per day and you will notice how incredible the results will be.

We want to read you!

The voice of our audience is very important for us. <u>We really appreciate to know your opinion about the book</u>, our main goal is the satisfaction of our users. We thank you for leaving your comment, better known in the community as: review.

¡Thank you for trusting us! New experiences always surprise you with good vibes:

Email us to **bluecherriespublishers@gmail.com** or go to **https://www.subscribepage.com/l4x8h2_copy2_copy** and you will receive a gift to continue stimulating your child in the process of development numbers skills.

1. Trace the numbers from 1 to 5 and count the pictures.

1   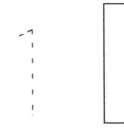

2   

3   

4   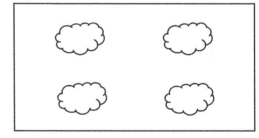

5   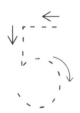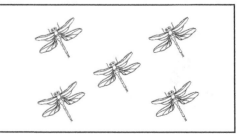

2. Trace the numbers 1-5. Then, draw your favorite picture according to the given quantity.

1   1   1

2   2   2

3   3   3

4   4   4

5  5   5

3. Count the dots on the mushroom and connect it to the number that is fitting to the quantity. Then, trace the number.

4. The worm Tilin. Trace the numbers, then, make papers balls. Glue them on the circles according to the given number and form the worm body. Tilin will eat the paper balls and his body will become longer each time.

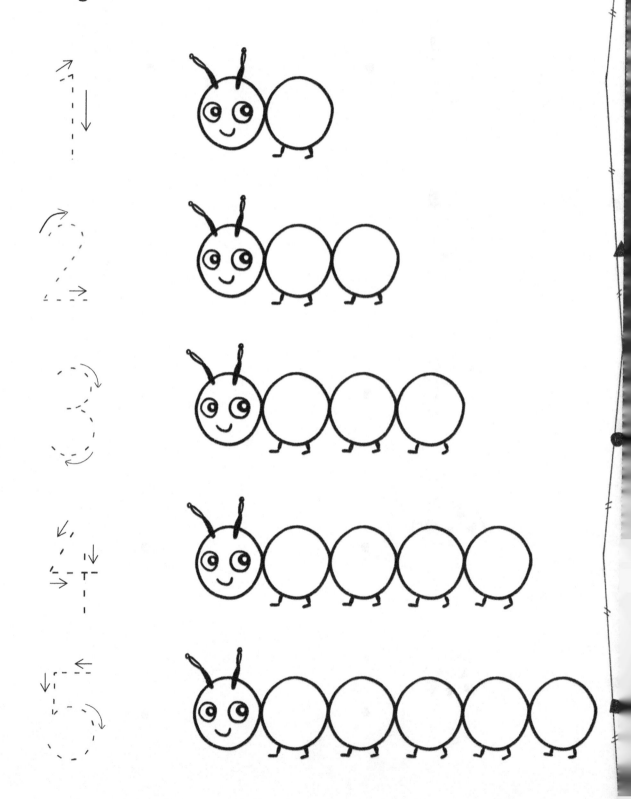

5. Draw with your favorite color the quantity of balls according to the given number.

2	5	1	3	4

6. Welcome numbers from 6 to 10. Trace the numbers and count the pictures.

6

7

8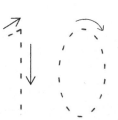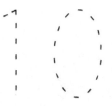

9

10

7. Trace the number, count and color the balls according to the number shows on each ice cream cone.

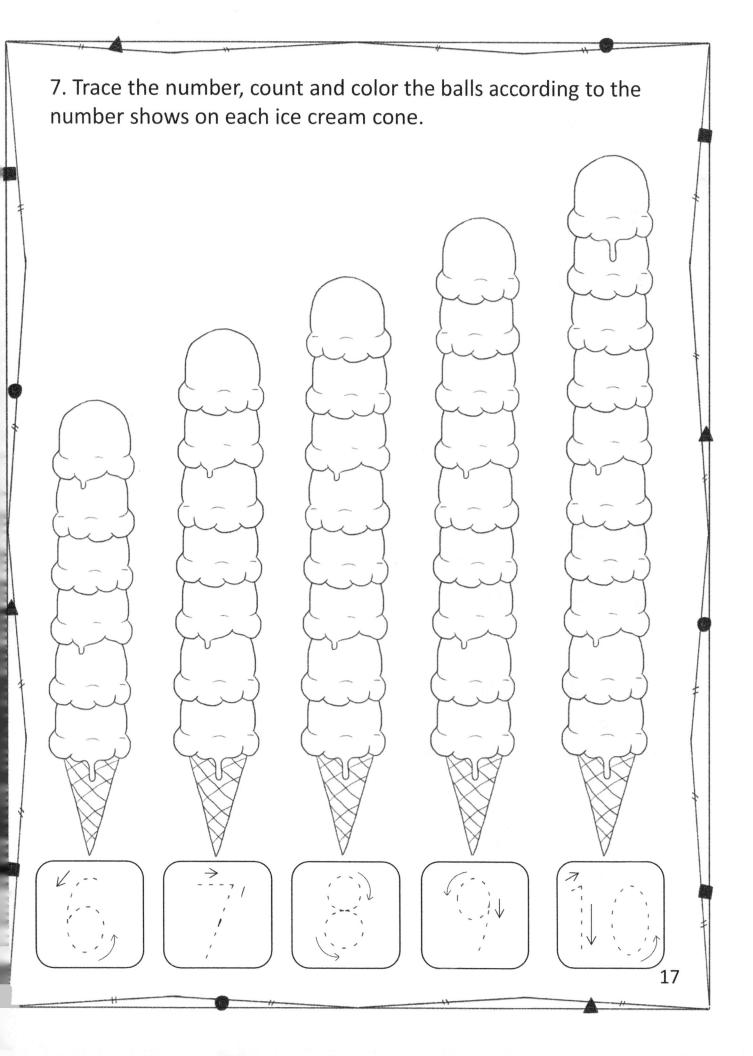

8. Count the dots on each die and trace the number that is fitting to the quantity.

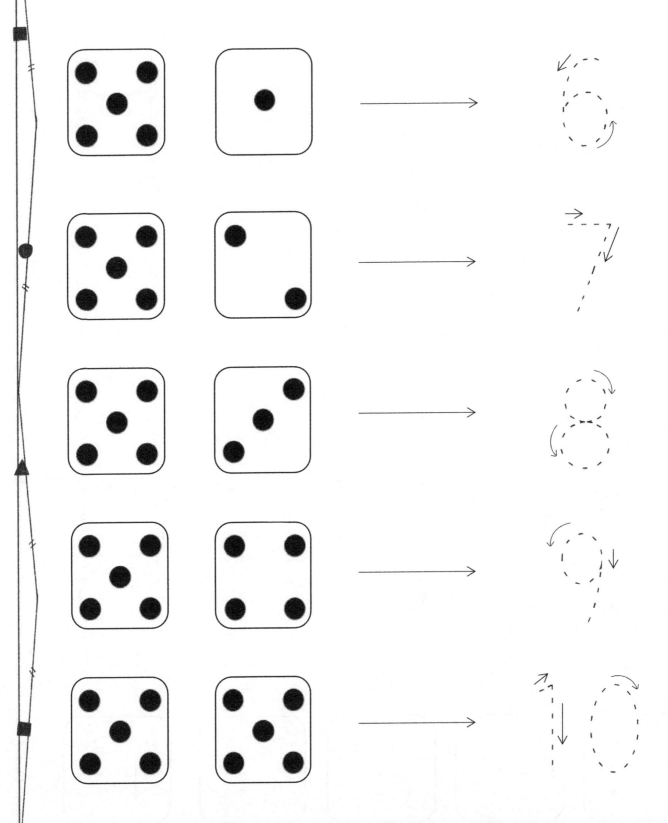

9. Numerical worm. Trace the numbers from 1 to 10, color the worm, then, cut it and fold it as an accordion. Your numerical worm will support you in the following activities.

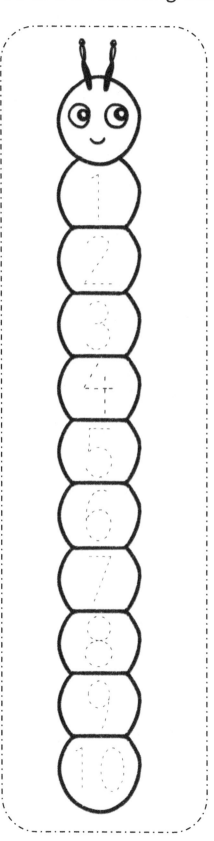

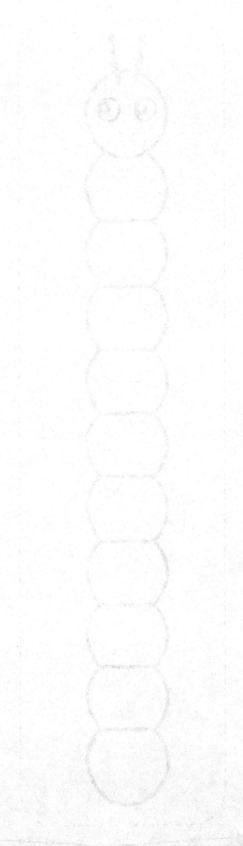

10. Checking the numbers from 1 to 10. Trace the number on each apple and color the correct quantity of squares according to the given number.

11. Trace the numbers on the worm and count them.

12. Connect the numbers from 1 to 10, complete the drawing and find out the hidden picture.

13. Count and trace the numbers on the ladybug. Notice how the ladybug will become bigger while you count.

14. Associate and color. Count the pictures and color the number that correspond to the correct quantity. Remember to use your numerical worm to do this activity

15. Connect the numbers to complete the fruit. Your numerical worm can also be with you during this activity. Color the picture and count the numerical sequence.

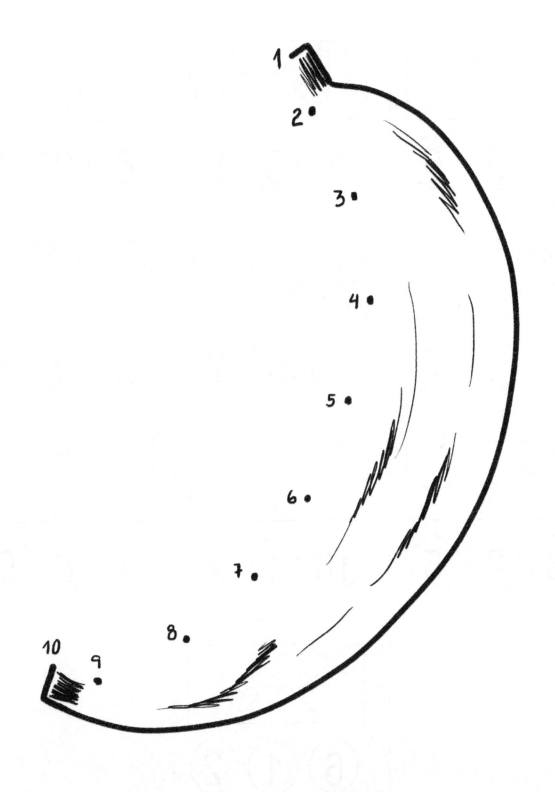

16. Counting and writing the numbers. Count the animals and color the circles according to the quantity. Then, write the correct number on the blank line.

17. Fishing, grouping and writing the numbers. How many animals of each species are in the aquarium? Write the correct number in the box next to each animal. You can use your numerical worm to remember how to write the numbers.

18. The numbers way. Complete the numerical sequence and help the bee to arrive to the flower and to the honey.

19. Count how many fruits are in the tree. Then, write the number according to the quantity that is fitting to each fruit.

20. Have fun tracing the numbers from 1 to 10.

1 2 3

4 5

6 7 8

9 10

21. Meeting the new numbers 11,12 and 13. Count the numbers to 10, then, trace the dotted numbers on the worm.

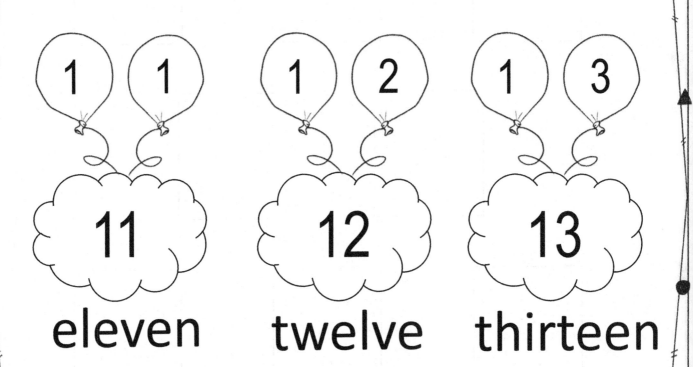

eleven   twelve   thirteen

22. Trace the numbers on the watermelon and color the quantity of squares according to the given number.

1 2 3 4 5 6 7 8 9 10

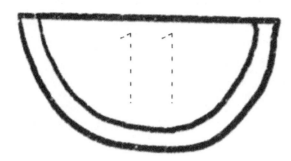

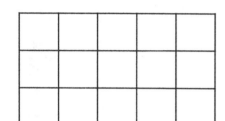

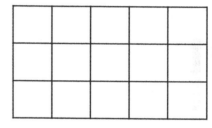

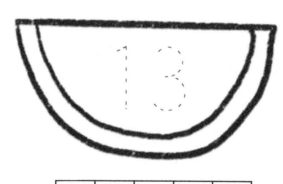

23. Count the numbers from 1 to 10. Trace the new numbers 11, 12 and 13. Then, follow each way and help the number 1 to meet their partners. Then, count the numbers that you connected.

1 2 3 4 5 6 7 8 9 10

11 12 13

1 _____ 1

1 _____ 2

1 _____ 3

24. Trace the numbers and draw the quantity of spots on each ladybug according to the given number.

1 2 3 4 5 6 7 8 9 10

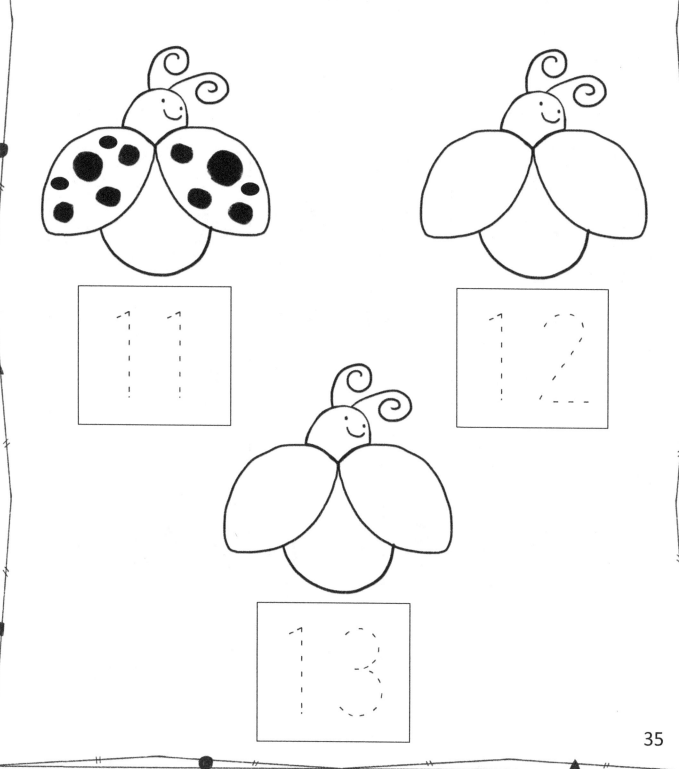

25. Count the drawing on each box and, according to the quantity, color the correct number.

1 2 3 4 5 6 7 8 9 10

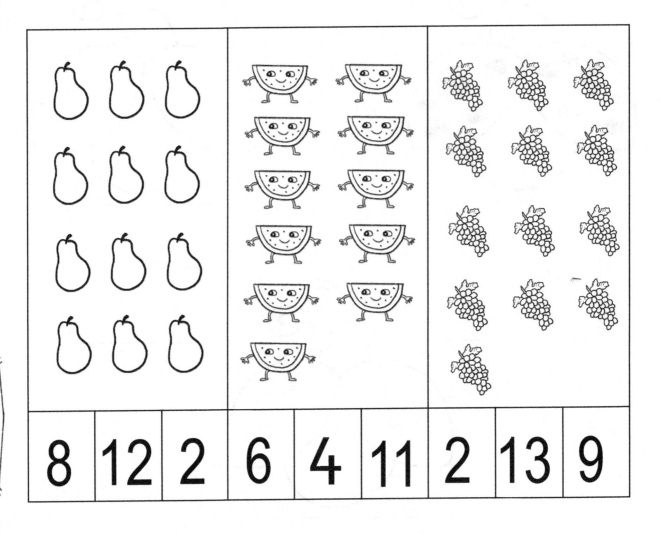

| 8 | 12 | 2 | 6 | 4 | 11 | 2 | 13 | 9 |

26. Learning the numbers 14,15 and 16. Help the frog to find the pond. Complete the numbers that are missing on his way. At the end, discover which are the new numbers and re-mark them.

[ ] → 2

5

[ ] [ ] → 9

11

13

14 fourteen   15 fiveteen   16 sixteen

27. Complete and trace the numbers on the worm.

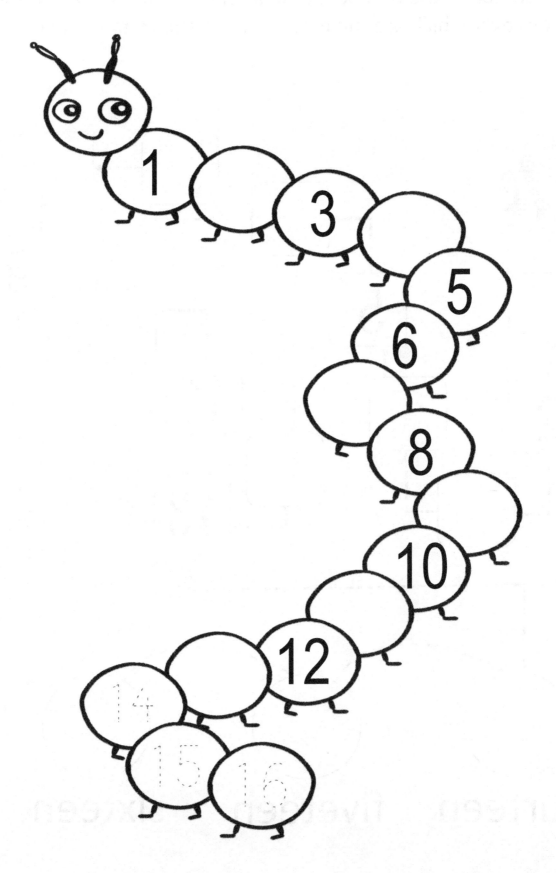

28. Count the numbers, connect them and discover the hidden picture.

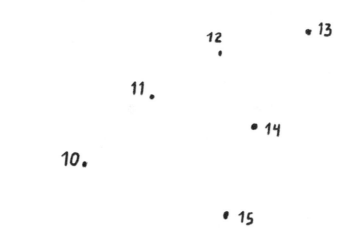

12 • 13

11.

• 14

10.

• 15

9 •

16

8 •

7 •

1

.2

6

.3

5

4

29. Count the pictures and connect them with the correct number according to the quantity. Trace the numbers.

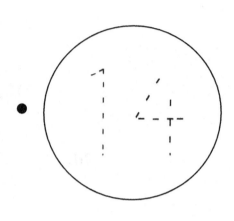

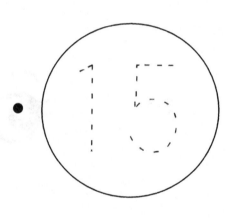

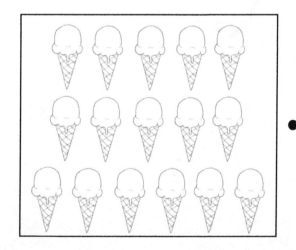
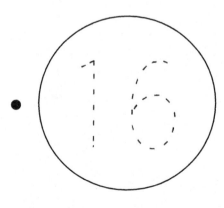

40

30. Count how many spots have the ladybugs and write the correct number between his wings.

31. Let's introduce the numbers 17, 18, 19 and 20. Help the mouse to eat the cheese. Count the numerical sequence, discover the new numbers and trace them.

seventeen

eighteen

nineteen

twenty

32. Identify the numbers on the moon and paint it according to the color that correspond to each number.

17 - Blue

18 - Gray

19 - Red

20 - Yellow

33. Count the objects and write next to it the number that is fitting to the quantity.

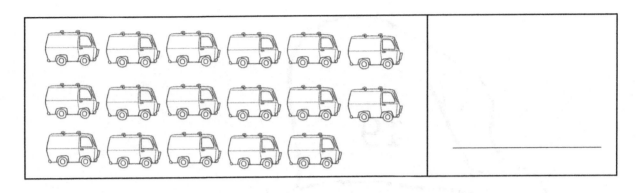

_____

_____

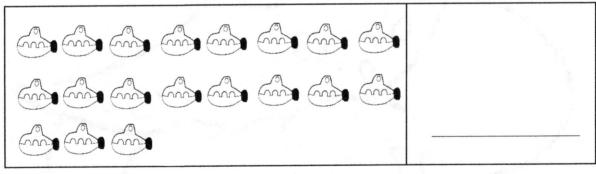

_____

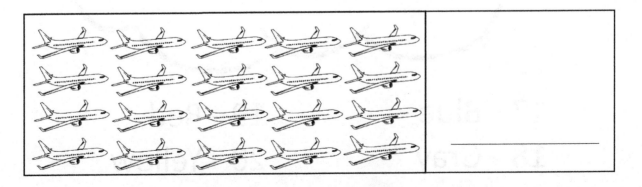

_____

34. Draw the way with a line from 1 to 20 and discover where the numerical sequence is.

1	13	17	18	19
2	7	16	2	20
3	11	15	14	4
4	5	20	13	12
18	6	1	3	11
10	7	8	9	10

35. Count the circles on each box and mark by using a check (✓) the correct number.

(8)	(12)	(20)		(3)	(16)	(7)

(6)	(18)	(2)		(10)	(5)	(9)

(4)	(1)	(11)		(4)	(12)	(1)

36. Count the pictures then mark with a check (✓) if the quantity correspond to the given number or with a (x) if the number is incorrected and, write the correct number on the blank line.

	Number	✓/✗	Correct Number
(9 squares)	9		
(20 stars)	12		
(14 suns)	14		
(15 smiley faces)	18		

37. Count and, then, write how many pictures there are.

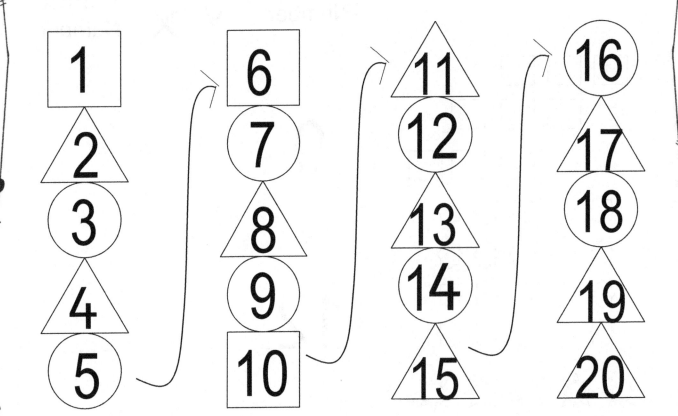

Write, How many pictures are there?

38. Count the spots on the mushroom and connect it with the correct number.

   •   •20

   •   •17

   •   •18

   •   •19

39. The number's pool. Look for the numbers that are swimming on the pool and answer the questions.

- ¿How many 12 are there? _____
- Circle the number 18.
- Trace the number 17.
- ¿How many 20 are there? _____

40. Identify the numbers on each apple, count how many boxes are marked, and connect the number with the correct quantity.

12 •

•

19 •

•

18 •

•

 20 •

•

41. Meeting the numbers 21, 22, 23, 24 y 25. Walk with the worm to find the apples. Trace the way and count the numbers that you already know. Then, discover the new numbers that are on the apples and color them.

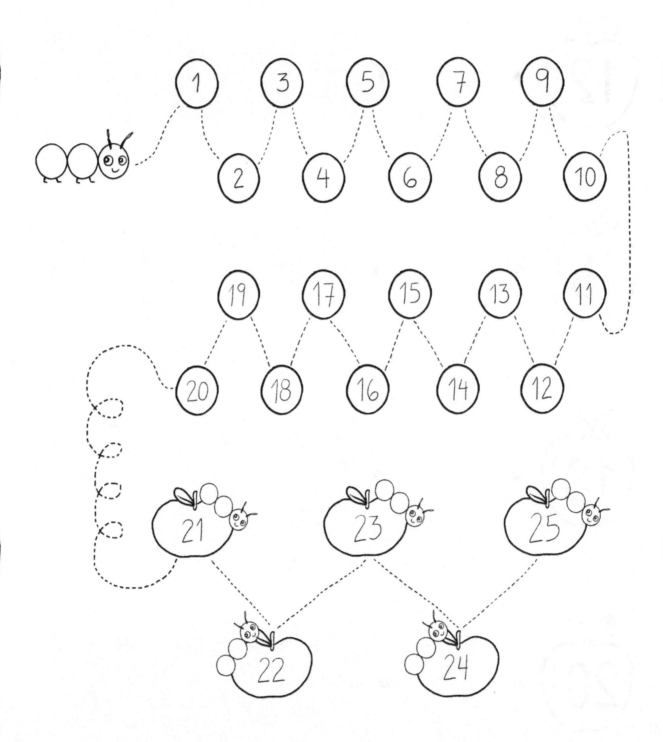

42. Number Maze. Follow the numerical sequence from 1 to 25 and lead the dog to his bone.

		5	16	15	8	1	3	8	7
		1	2	18	20	9	4	7	6
		22	3	10	6	15	3	16	8
3	10	18	4	16	1	2	20	21	22
15	7	6	5	25	17	18	19	1	23
23	8	3	1	7	16	1	7	11	24
1	9	10	2	14	15	21	8	9	25
13	2	11	12	13	7	3			
16	17	20	19	5	8	12			

43. Trace the numbers, then count the pictures on each square and connect them to the correct number.

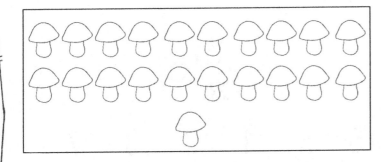 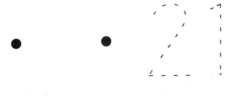

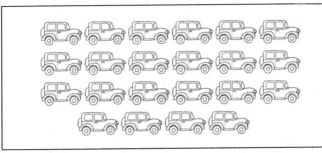 

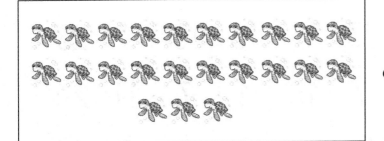 

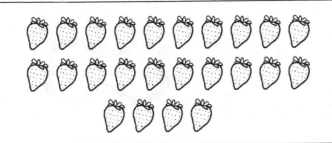 

44. Numbers dictation. Cut the numbers and spread them out on the table. Listen to the dictation and glue each number on the blank line. Numbers to be dictated: 23, 16, 18, 11, 25, 12, 21, 14, 19, 22.

a. _____ _____   b. _____ _____   c. _____ _____

d. _____ _____   e. _____ _____   f. _____ _____

g. _____ _____   h. _____ _____   i. _____ _____

j. _____ _____

- -

1	1	1	1	1	1	1
1	2	2	2	2	2	2
3	5	6	8	4	9	

45. Count carefully and answer the question: How many triangles, circles and squares are there?

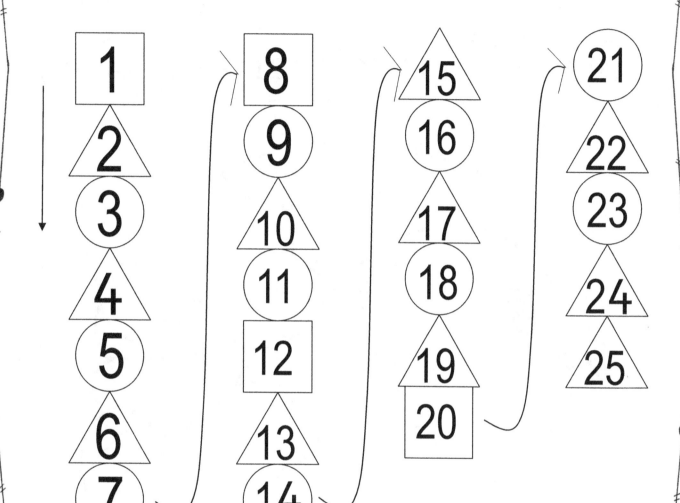

How many triangles, circles and squares are there?

 _____  _____  _____

46. Let's welcome the numbers 26, 27, 28, 29 and 30. Count the numbers that you already know and color the fish with the new numbers.

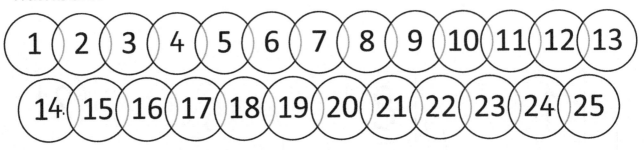

26 - Blue   27- Red   30 - Green
28 - Orange   29 - Yellow

47. Count the numbers and trace those that the crocodile ate.

| 1 | 2 | 3 | 4 | 5 | 6 | 7 | 8 | 9 | 10 |

| 11 | 12 | 13 | 14 | 15 | 16 | 17 | 18 | 19 | 20 |

| 21 | 22 | 23 | 24 | 25 |

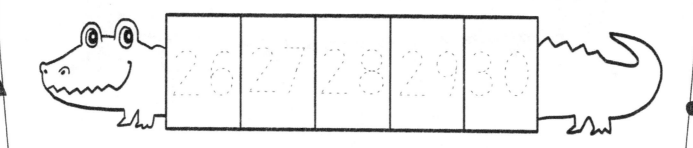

48. Decorate the party and trace the new numbers that you learned.

49. Count and connect the numerical sequence from 1 to 30 and discover in which fruit the worm lives.

28

29

30

27

1

26

2

25

3

24

4

23

5

22

6

21

7

20

8

19

9

18

10

17

14

13

11

16

15

12

50. Count the numbers and write those that are missing to help the ship to arrive to the island.

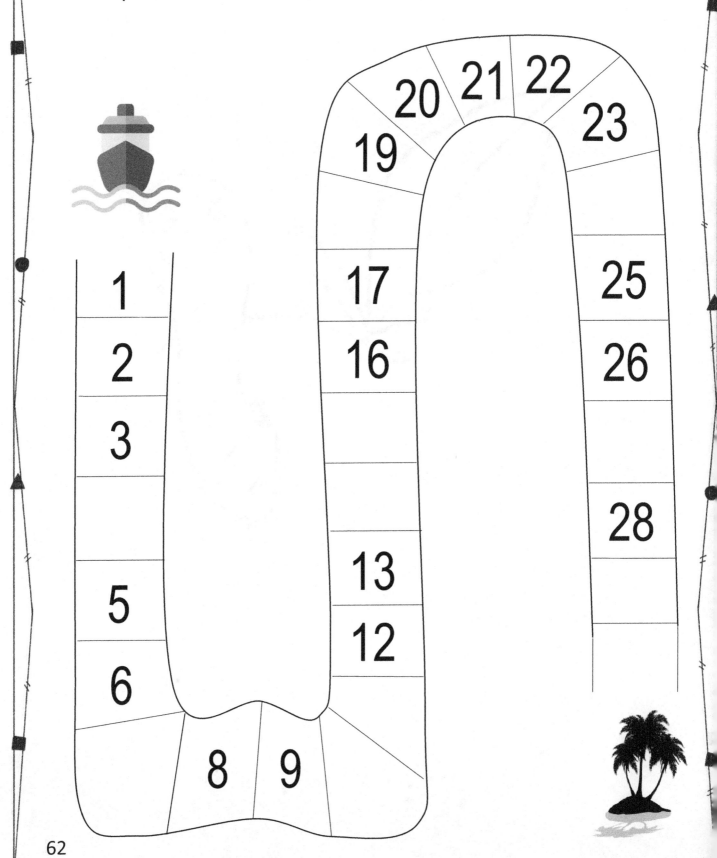

Conclusion

The activities of this book have a cheerful, unique and funny content that not just support the kid to learn the numbers but also to find a way to be connected with himself or herself and with the magic of the numerical world. The kid will begin the journey of the number knowledge in a satisfactory way, and at the same time, will flow in the learning process.

The success to learn something new is directly proportional to the tools given to the kids, as well as, the fact to let them to know themselves by doing the activities. The education is important and its effects are more significant if it is applied in a healthy and conscious way.

Help other children to live the experience of development the numbers skill, and recommend the activities in this book

with dedication&love
Rosi and Coni

Made in the USA
Monee, IL
13 September 2024

65750149R00037